The Food Additives Dictionary

by **Melvin A. Benarde, Ph.D.**

Professor and Chairman, Department of
Community Medicine and Environmental
Health, Hahnemann Medical College and
Hospital

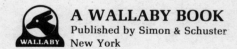

A WALLABY BOOK
Published by Simon & Schuster
New York

Copyright © 1981 by Melvin A. Benarde

All rights reserved
including the right of reproduction
in whole or in part in any form
Published by Wallaby Books
A Simon & Schuster Division of Gulf & Western Corporation
Simon & Schuster Building
1230 Avenue of the Americas
New York, New York 10020

WALLABY and colophon are registered trademarks
of Simon & Schuster

Designed by Irving Perkins Associates

Manufactured in the United States of America

Printed and bound by The Murray Printing Company

1 2 3 4 5 6 7 8 9 10

To Dana
Andrea
Scott
Anita

And to those who walk on eggs,
rather than enjoy them

Contents

Introduction

Recently I had the good fortune of being a participant on a TV talk show dealing with food additives. About midway along, a woman called to say, "We've got all these additives on food labels—what do they mean?"

The light clicked on: "That's it! People want to get beyond the incomprehensible names. They want to know what it is they are eating." Of course, they *should* want to know.

When I arrived at my office the following morning my secretary said, "Well, are you going to do it?" "Do what?" I asked. "What the woman asked for last night. Explain how to know the additives." "Of course we're going to do it." For a moment I was afraid she knew me better than I knew myself.

My first step was to find out which additives people were concerned about. I did this by conducting a survey. This book is the result of a survey I conducted of people from Philadelphia, Trenton, Princeton, and New York—a sample quite similar to the general population.

Traveling by train as I do each day between my home in Princeton, New Jersey, and Hahnemann Medical College and Hospital in Philadelphia, I was able to ask a number of commuters, and others, if they had questions or were worried about food additives. If they said they did, and many did, I asked them to write the name or names of any ingredient they were concerned about, on cards I supplied. In this way, I amassed several hundred responses.

There were, of course, many duplications. Almost everyone was concerned about BHA and BHT, nitrates and nitrites. This book thus deals with those additives real people wanted to know more about.

Since food additives are a fact of life, a part of our food supply, my object in preparing this book was to help answer the queries, What do they mean? What are they? and Are they safe? I also hoped to bring a modicum of light to a subject fraught with heat. My goal will have been fulfilled if people refer to the book as they shop.

1

Additives:
A Brief Primer

A public dependent upon a vast industry for its food ought not to have to worry about the safety of the supply—it should be taken for granted. For a long time it was. Unfortunately, that has changed. Over the past fifteen years, our food has become an object of fear: fear that cancer lurks in every loaf of bread, breakfast cereal, and cake mix—in any food containing additives, chemicals that "aren't natural."

Clearly, *fear of cancer, not cancer* has reached epidemic proportions in the United States. Oncophobia is abroad in the land. That the best available epidemiological and clinical evidence points away from food additives as causes of cancer and toward cigarette smoking, alcohol consumption, genetic predisposition, lack of exercise, and overeating, appears of little concern to those who prefer to believe that food additives must be harmful.

The idea that our food supply is safe and nutritious seems to be unbearable, and is greeted with ridicule and contempt. As a consequence, misrepresentation, misinformation, and their progeny, fear and confusion, are the public's lot.

There is an almost wanton disregard of the available evidence supporting the safety of food additives. A recent conference on primary cancer prevention held in New York in June, 1979 made a point of noting the almost total lack of evidence for food additives as carcinogens. Neither nitrate, nitrite, nor any other currently used food additive is known to cause cancer in man. Although experiments have shown that certain nitrosamines can cause cancer in animals, direct association between nitrosamines and human cancer has not yet been found. It may be that our general cynicism taints our view of our food supply.

Considering the volumes written and spoken about additives, it is surprising how little people really know about them.

Nevertheless, additives are here to stay. And as the public requested, and now requires, their names adorn all food labels. They have become part of our world. It's not unreasonable, then, to want to become familiar with them—to want to demystify them. They should become household words devoid of "fear and trembling."

WHAT ARE ADDITIVES AND WHY DO WE USE THEM?

What, in fact, do these barely pronounceable cognomina tell us about themselves? Stripped of their congressional jargon and legalese, *additives are ingredients, substances, chemicals, intentionally added to food to produce specific effects.*

Although we like to think we humans are unique, it may come as a surprise to learn that chicken, steaks, bread, cheese, franks and beans, fish, shrimp, ham and eggs, peanut butter, jam and jelly, and just about anything else you can think of are as compelling and nutritious to microbes—bacteria, molds, and yeasts—as they are to us. In fact as some wag has said: "Life is nothing more than a race between man and microbes to see who gets the food supply first."

The protein, carbohydrates, fats, trace minerals, and vitamins in our food are as necessary and vital to the growth and development of bacteria, insects, rats, mice, and other scavengers as they are to us. The yearly loss of food to pests of one kind or another is staggering. With all our high technology, we lose some $2 million of food to insects. The loss to microbes is similarly astronomical. But that isn't all. Rancidity and other off-odors and tastes result from the natural processes—the chemical reactions—that occur in such complex chemical systems as potato chips, fish fillets, meatballs, chicken cacciatore, crackers, cookies, and cake—these things which we call food, but which, in fact, are nothing more than a variety of chemicals naturally strung together in unique patterns. It may be novel, but you can think of it that way.

To protect our foods from a host of plunderers, we use a variety of *preservatives*—food additives.

Another group of chemical additives are used to keep fats and oils from becoming rancid. Propyl gallate, butylated hydroxyanisole, and butylated hydroxytoluene, mercifully shortened to BHA and BHT, are representatives of this group.

An assortment of chemicals are available from which to choose an appropriate one for an appropriate food—to keep a food from drying, or to impart desirable crispness, or to maintain stable mixtures of liquids that would simply not mix or would separate quickly—we eat these in chocolate, mayonnaise, and peanut butter, as *emulsifiers*—food additives.

Soups, sauces, and syrups, for example, contain sta-

bilizers and *thickeners* to give desirable viscosity and texture—and to prevent sloppy puddings.

And there are *sequestrants* to combine with trace metals—metals that can produce off- odors and tastes. And there are *solvents* and *propellants* and *sweeteners*, and *humectants* to retain moisture; and of course, there are flavor enhancers and coloring ingredients and anti-gushing chemicals, which prevent beer, for example, which is under great pressure, from gushing when cans or bottles are opened. Chemicals are added to achieve a dozen different effects in different types of foods.

But that's only part of the story.

Contradictory as it may seem, we live in a world of double paycheck families in which *neither* breadwinner has the time or inclination to spend long hours in food preparation, but both want scintillating meals. We also live in a world of more and more free time—time we prefer to use for recreation and education rather than in the kitchen. Ready-to-eat foods are a necessity in a leisure-type world. And that means additives.

Sad as it may seem, we no longer live in a world where we can each supply our own needs from the family garden or farm. Our foods come to us not only from border to border and "sea to shining sea," but from Europe, the Middle East, Africa, Asia, and South America. We literally demand an extraordinary variety of food—year round, never mind the season.

And we are securely tied to a huge food industry that must supply food to some 220 million of us. This food has to be farmed, fished, prepared, packaged, stored, shipped, and ready to be eaten when sold or when taken from shelves at home after we buy it, no matter how long it sits there.

But these reasons for using additives do not complete the story.

Some foods require fortification, others refortification. A number of foods are naturally deficient in essential nutrients, while others lose a portion of their natural content during preparation and manufacture. In these instances, additives are used to either return the food to its original nutritional level or bring it up to a more desirable level.

Nutrient additions to food have, in fact, proved to be an effective way to prevent nutritional deficiency diseases. The addition of vitamin D to milk is a good example. As a direct result of the addition of small amounts of vitamin D to milk over the past fifty years, rickets, once a common illness, has been virtually eliminated. Pellagra—once the scourge of the South—with its dermatitis, diarrhea, dementia, and death, is a thing of the past because of the simple addition of niacin, a B-vitamin, to those foods naturally deficient. This is a success story. But that's not all.

Food processors as a group appear to believe that most people prefer appropriately colored food. Red meat is preferred over shades of gray, and orange oranges are preferred over green and brown. This is called "appeal."

Consequently, to make foods more appealing, color becomes an integral part of what it takes to make foods not just edible, but desirable.

BUT ARE THEY SAFE?

Apparently, then, it is reasonable that additives become ingredients in food. But are they safe? And perhaps just as important, what do we mean by "safe"?

I raise this seemingly simplistic and self-explanatory question simply because it is *not* simple. But neither is it so convoluted nor abstruse as to be beyond consideration. I suspect that much of the confusion about the nature of safety decisions would be dispelled if the meaning of "safety" were understood.

Before we can define "safe" or "safety," the specter of risk must be raised, and along with it the question, Are we willing to make choices? Of course, we are. We do it every day, often many times a day. Inherent in choice is concern for risk. Are we willing to take risks? Again, of course we are, but quite patently not any and all risks. Well, then, *acceptable risks.*

If we can go that far, we can define "safety" as a judgment—that's the key, a personal *judgment*—about the acceptability of risk, and in turn we can define "risk" as a measure of the probability of harm to our health. I'm saying that safety is a value judgment of how much risk we—as individuals, or collectively as a society—will accept, and that risk is a quantitative assessment of the degree of harm to health that may be anticipated. Consequently, a thing, a chemical, a food additive, a drive to the beach, or a plane ride from Chicago to Miami is safe if its risks are judged to be acceptable—or unsafe if they are judged to be unacceptable.

Obviously, this definition is not the same as that in most dictionaries, which define "safe" as "free from risk"—a definition both unacceptable and misleading. Nothing can be totally without risk—not even taking a bath. There are only *degrees* of risk, and thereby, degrees of safety.

This concept emphasizes the relativity and judgmental nature of the concept of safety. In this system, there can be no absolutes; a thing can never be either all black or all white. That's terribly important. Shades of gray are, in fact, nature's way.

This definition implies that there are two very different aspects in determining how safe things, including food additives, are: (1) *measuring* risk, a technical or scientific problem, and (2) *judging* the acceptability of that risk, a matter of personal and social values.

Failure to appreciate this dichotomy may just be the shoal upon which understanding founders. Failure to appreciate this duality gives rise to the false expectation that scientists can *measure* whether something is safe. They can't. They are limited to the first function—measurement of probabilities and consequences, deaths, injuries, or disease. The second, benefit or value to people, is *not* a decision scientists should make or be asked to make. Deciding whether people might or should be willing to *bear* the estimated risks is a "judgment call" that scientists are no more qualified to make than any other citizen.

Safety, as we have all come to realize, is a relative attribute that changes from time to time and place to place. Examples of this are legion. And society—people—clearly shows its acceptance and rejection of risks, though it may not overtly think of its behavior in these terms.

Deaths from automobile accidents, for example, rose to 51,676 in 1980—an increase of 1.2 percent over 1979. And there is no reason to expect or anticipate a decrease in the carnage on our highways and streets in the years ahead. Automobiles, motorcycles, people, and too often alcohol, combine to raise deaths from accidents to the fourth leading cause of death in the United States today.

It has recently been estimated that 2,000 additional deaths over the next twenty years—100 extra deaths per year—*may* be related to radiation effects from all our nuclear reactors. If the public outcry against radiation is any bellwether, a large segment of the public is unalterably opposed to such a risk. Choices are indeed made and risks taken. More than 51,000 deaths per year from motor vehicles is quite obviously acceptable; 100 deaths per year from radiation is not. "Safe," then, is anything people decide is safe.

THE FDA WAY

This is one way of looking at safety. The Food and Drug Administration has another approach. For them, "safe" or "safety" means that there is "reasonable certainty in the minds of competent scientists that [in the case of food additives] the substance is not harmful for the intended conditions of use."

Considering the utter impossibility of establishing with certainty the absolute harmlessness of *any* substance, the FDA uses three factors or guidelines to establish safety of food additives:

1. The cumulative effect of a substance in the diet, taking into account any chemically or pharmacologically related substance in the diet. The issue here is: do increasing amounts of chemical taken over extended periods become hazardous? Included in the total arithmetic are naturally occurring chemicals whose molecular structures are similar to that of the food additive.

2. The probable consumption of the substance and of any other substance formed in or on the food, because of its use. Here the concern is that one plus one may not equal two. Some chemicals are naturally

more outgoing than others—they are less stable. Hence, they are quick to enter into new relationships. Nitrite is a good example. It is readily absorbed from the intestinal tract and rapidly disappears from the bloodstream. At times, however, in the presence of certain amino acids, a new substance is formed—nitrosamine. And, where the original nitrite was innocuous, the new compound may not be. It surely *may* be, but this remains to be ascertained.

3. Safety factors which, in the opinion of experts qualified to evaluate safety, are generally recognized as appropriate.

As a consequence of these guidelines, two categories of *safe* food additives have been established. One is the *Generally Recognized As Safe* or GRAS list*, which contains "any substance of natural biological origin that has been widely consumed in the U.S. prior to 1958, without known detrimental effect, for which no known health hazard is known."

A second category contains those ingredients that are not GRAS. These require that evidence of safety be demonstrated—by those who want to have the substance become part of the food formulation. The onus is on those who want the chemical used to show that it is harmless *before* it is used in food.

The type of evidence required and most often submitted consists of results of both short- (three to six

---

* New information may at any time require reconsideration of the GRAS status of an ingredient. An example of this is the recent decision to remove caffeine from the GRAS list and place it on the Interim List, which requires that additional information on its toxicity or lack thereof be obtained.

months) and long-term (two years) animal-feeding studies.

Should FDA scientists—or outside consultants hired by them to evaluate submitted petitions—find the data acceptable, a food additive regulation is drawn up and circulated—usually in the Federal Record. Food additives for which a regulation has been promulgated are known as "Cleared."

If an ingredient is neither GRAS nor Cleared, it should be rejected.

INTRO TO FOOD FOR SOME THOUGHTS

Writing about his friend, Ford Maddox Ford the literary critic, Ezra Pound said, "I once told Fordie that if he were placed naked and alone in a room without furniture, I would come back and find total confusion." Total confusion may well be the most appropriate description of the public's understanding of a number of essential biochemical concepts.

Ideas such as organic, synthetic, natural and artificial appear to be subjects of great concern to large segments of the eating public, but are no less subjected to great contortion. Consequently, early on in this book, I relish the idea of offering some food—for thoughts on these strongly held notions.

2

Food for Some Thoughts

Organic. On page 1014 of the Random House Dictionary of The English Language (Unabridged, © 1967, 1966 by Random House,) fifteen listings are given for "organic." The first notes that "organic" formerly pertained to those compounds—substances—derived from plants and animals, but now includes all compounds containing the element carbon.

In its introductory chapters, just about every textbook of organic chemistry and/or biochemistry asks the question, What is organic chemistry? The answer is brief and straightforward: the study of carbon compounds.

Although all organic compounds have carbon as their structural unit, hydrogen too, is always present. Often present are oxygen, nitrogen, sulfur, and potassium. Other elements such as iron and cobalt may be present.

By 1975, over a million organic compounds had been cataloged. Ninety percent were synthetic; that is, they were man-made. The remaining 10 percent were obtained from animals, higher plants, and microorganisms (lower plants). Most of those 90 percent are indistinguishable from their naturally occurring counterparts. The synthesis of naturally occurring chemicals means that their availability is no longer subject to the vagaries of weather and political upheavals, they do not vary from batch to batch, and they are often less expensive. Ergo, the interest in synthesizing chemicals.

The naturally occurring compounds include proteins, carbohydrates, fats, vitamins, and hormones, while the synthetic are derived primarily from natural sources of carbon—coal and petroleum. Only proteins and certain complex carbohydrates such as lignin have yet to be synthesized by chemists in their laboratories.

Our world can be divided into organic and inorganic things. All things are either compounds of carbon or they are not. Children, clothing, chairs (wooden), chrysanthemums, and cauliflower are organic. Table salt, sodium chloride, is inorganic—no carbon, no hydrogen. And so are lead and the chlorine we put in our swimming pools. And water too, even though it contains hydrogen. It's the carbon that makes the difference.

It is terribly important to reiterate that although naturally occurring carbon compounds can be made in a laboratory, it is extremely difficult, even impossible, to distinguish between them—biologically or chemically. Vitamin A, whose real name is 3, 7-dimethyl-9 (2,6,6,-trimethyl-1-cyclohexen-1-yl)-2,4,6,8-nonatetraen-1-ol (What's in a name, you may ask!), whether extracted from fish oil or egg yolk (the vitamin is not commercially obtained from plants) or constructed in a laboratory, are identical products—and with the same activity and potency, unit for unit.

The vitamin D_2 added to milk is synthetic. Years ago, when I was a kid, calciferol, another name for the foultasting stuff, was straight fish liver oil. Try swallowing that sometime. Currently, it is far too expensive to harvest oil from fish livers. What to do? A plant sterol, ergosterol, is subjected to ultraviolet irradiation and vitamin D_2 is produced.

There is another way to get adequate quantities of vitamin D. If only I'd known it when my mother said, "Open."

D_3 is another of the several chemically related compounds called vitamin D. D_3 is a derivative of cholesterol—7 dehydrocholesterol—which is found in the skin. That's right. When the skin is exposed to sun (ultraviolet, or UV light) sufficient vitamin is formed to protect against rickets. All that's needed is to be sure you live in the right place. Can't you just hear Gertrude Stein summing it up—a vitamin is a vitamin is a vitamin and a chemical is a chemical. Of course they're the same, whatever their origin.

Artificial. According to my favorite dictionary, artificial means "made by human skill," or "produced by man as opposed to nature." Does that include chicken soup?

Natural. This refers to those things—including chemicals—"existing or formed by nature, as opposed to artificial." The chicken for chicken soup?

Synthetic. "Pertains to substances formed by chemical reaction in a laboratory, as opposed to those of natural origin." Unfortunately, this definition does not cover making a cup of coffee, tea, or cocoa. The simple act of boiling or steeping organic material—tea leaves, herbs, coffee, cocoa, or chicken, as in chicken soup—releases a host of organic chemicals. It's these organics bubbling about on the surface whence arises the tantalizing aroma we recognize as chicken soup. Just as it is the organics in hot coffee and apple pie that draw us to the kitchen. Cooking is just one chemical reaction after another. Synthetic? Artificial? Natural?

Consider what the Federal Trade Commission (FTC) has said about the benefits of "natural" foods.* For the record, they noted that,

> It must be regarded as scientifically established that natural foods are not inherently superior either in terms of nutrient content or safety to foods that are not so characterized.

They went on to say that,

> In many instances natural foods may not have as high a nutrient content or be as safe as "unnatural" foods which have been fortified and are more highly processed.

These words appear to have been forgotten or quite simply overlooked.

But that wasn't all. The folks at the FTC continued in this same vein by noting that it was clear from the available evidence that,

> . . . There is no basis whatsoever for a superiority claim merely because a food is "natural." Thus the claim that a natural food is inherently superior to its non-natural counterpart is both false and unsubstantiated.

> There is no difference between organic and inorganic** food, thus leading to the conclusion that

---

* 16 CFR, Part 437, Phase 1, Nov. 29, 1978 (Note: "CFR" stands for the Code of Federal Regulations.)

** Here the commissioner was not referring to the inorganic, non-carbon-containing substances. He was rather making a distinction between the foods termed "organic" by their purveyors and foods sold as nothing more than food in the usual food stores. Of course, food by its very nature is organic.

the distinction drawn is spurious from a scientific standpoint.

In concluding their testimony, they said that,

The vast preponderance of the evidence indicates that organic foods are not nutritionally superior to ordinary food.

I can just hear him—Josh Billings. This teapot-tempest might well have elicited one of his famous lines: "The best bill of fare I know of is a good appetite."

3

Fresh—What's Fresh?

Pluck an apple from a tree, or a peach or pear. That's fresh. Fresh is also an ear of corn that you pick, shuck, cook, and eat. Must the water be boiled and waiting for the ears? Tomatoes and cucumbers picked from a garden just prior to dinner must surely be fresh.

Fruits and vegetables brought to canneries and refrigeration plants are often processed, canned, or frozen within hours of being harvested.

Store-bought fruits and vegetables grown around the state and country can be weeks old before being purchased and eaten. Are they "fresh" produce?

Kept properly iced, "fresh fish" can be a day to a week old before it's cooked and eaten. And what of that moist, light, chocolaty cake made from a cake mix? Is that fresh?

Given these examples, what can or should we call "fresh"? How much time from harvesting, catching, butchering, or canning to eating may elapse for a food to be still considered fresh?

Unfortunately, dictionaries are of little help, as they define "fresh" as being "newly obtained," "not canned or frozen." Considering the speed with which today's foods are picked, caught, frozen, canned, or baked, can they be as fresh as "fresh" food?

Unfortunately or fortunately, the foods we've traditionally called "fresh" do not carry nutritional labeling. Consequently, at time of purchase, we know nothing of their nutritional content. Are they higher or lower in protein, carbohydrates, fats, vitamins, and minerals than their canned or frozen relatives? Do our eyes perhaps deceive us? It's something to think about.

4

Color in Your Life

Under the Delaney Clause of the Food Drug and Cosmetic Act, any food-coloring ingredient which produces cancer in laboratory animals, *in any amount*, may not be used in food. The ban is absolute.

As a consequence of this stringent criterion, only seven synthetic colors have been certified for use in food—just seven:

Blue No. 1
Blue No. 2
Yellow No. 5
Yellow No. 6
Red No. 3
Red No. 40
Green No. 3

Two others, Citrus Red No. 2 and Orange B, are permitted but with narrow limitations.

Citrus Red may be used only to color oranges, and specifically those not intended for further processing, while Orange B is allowed at a specified level and may be used only for coloring the casings and surfaces of hot dogs and sausages. At this writing, however, the sole manufacturer of Orange B has withdrawn it from sale.

Furthermore, although the seven have been certified harmless, they actually have a provisional status—just as all additives do. This provisional status means that they can be withdrawn at any time that new data indicate lack of safety.

How is it possible for seven colors to satisfy the needs of the thousands of foods offered for sale? They can't. Quite obviously, more than the seven certified synthetic colors are used. The fact is that most colors used in food are drawn from the more than twenty-odd natural extracts that have intense coloring ability and are generally recognized as safe for human consumption. (See page 30.)

This business of natural versus synthetic has its humorous and baffling side. For example, the chemical that gives tomatoes and carrots their red-orange color is carotene. Carotene, of course, is widely distributed in nature, occurring in a host of plants. Not too many years ago, chemists were able to duplicate the molecular structure of carotene in the laboratory. The natural and the laboratory versions are so alike that, much like single-ovum twins, they can't be told apart.

In the United States, if food processors want to add carotene to a food, and if they want to use the laboratory-made twin, which is cheaper, more readily available, and always of the same potency, the FDA requires that the food label carry the designation "artificial color." Across the border in Canada, the same colorant must be labeled "natural color."

Do we need color in food? I suspect that most people would respond with a resounding yes. Color is important because it makes food appealing. Color helps us enjoy our meals. Food is more than just something to satisfy the body's physiological needs—for chemicals. And no, color isn't necessary for survival, but eating must surely be more than merely a subsistence activity.

COLORING INGREDIENTS
Synthetic Pigments

FD&C	Blue No. 1
FD&C	Blue No. 2 *
FD&C	Red No. 40
FD&C	Green No. 3 *
FD&C	Red No. 3
FD&C	Yellow No. 5
FD&C	Yellow No. 6 *
FD&C	Citrus No. 2
FD&C	Orange B

* Provisional until additional data permit certification. Although January, 1981, was their review date, it has been extended, with no set date as of June, 1981.

COLORING INGREDIENTS
Natural Products

Annatto
Beet powder (dehydrated)
Beta-Apo-8-carotenol
Beta-Carotene
Canthaxanthin
Carminic acid
Corn endosperm oil
Carrot oil
Ferrous gluconate
Grape skin extract
Iron oxide
Paprika and oleoresin
Riboflavin
Saffron
Tagetes (aztec marigold) meal and extract
Titanium
Toasted partially defatted cooked cottonseed flour
Turmeric
Vegetable juice

These have been judged safe and do not require any special clearance.

5

What's the "Junk" in Junk Food?

On any day of the week, articles of one stripe or another warn us away from eggs, cream (both sweet and sour), a host of cheeses, meat, animal fats, cured meats, bacon, smoked meats and fish, sandwich meats, salami, bologna, and organ meats—particularly pancreas, sweetbreads (thymus), and kidney. This litany is not exhaustive. Nevertheless, we are warned away from all these foods as a means, we are told, of avoiding heart disease and cancer.

As if that list of "lost" foods were not long enough, stentorian voices have lately been trumpeting the horrors of "junk" food—hot dogs, potato chips, candy, soda, french fries, and breakfast cereals, to note a sampling. We seem to be left with nuts and beans.

Attempts, some of them successful, have been made to bar or remove candy and "snack" foods from schools around the country—again, in the name of protecting the health of the public. If some folks have their way, "snack" or "junk" foods—terms used interchangeably—are on their way to becoming the next pariahs.

But what, in fact, is the "junk" in junk food? Or to

put it another way, is "junk" food really junk?

Rushing to my dictionary, I find that "junk" is slang for trash or rubbish, and that trash and rubbish are worthless things—refuse, garbage to be gotten rid of.

Considering that I was brought up on frozen Milky Ways, Tootsie Rolls, and chocolate-covered raisins, and still find it difficult not to devour a bag of potato chips at one sitting, the idea that some of my favorite snacks—foods—may be worthless junk, evokes a degree of umbrage—enough at least to make me want to explore the issue.

Looking at the wrapper, I find that my Milky Way contains milk chocolate, corn syrup, sugar, milk, vegetable oil, salt, malt, and egg white. Are these trashy or unhealthy? They sure taste good.

As I understand these ingredients, chocolate comes from cocoa beans, a natural product grown on trees. And egg whites are a naturally occurring protein without a hint of cholesterol—if that's a problem. Corn syrup comes from corn, which is another natural product, as are milk, an almost complete food, and sugar from sugar beets or cane, a pure carbohydrate.

My 47-gram bar—1¾ ounces—two to three bites, if no one is around to share it—supplies 210 calories, 6 percent of the Recommended Daily Allowance (RDA) of both calcium and iron, and 2 percent of the RDA of protein. It's really quite nutritious. Hardly something to be discarded as refuse.

In addition to chocolate and milk, my Tootsie Roll contains whey, a biologically available protein, soybean oil, and vegetable lecithin. Lecithin appears to be a highly desirable substance, if one can go by all the advertisements for it, as well as its inclusion in a variety of hand lotions and face and body creams. My Tootsie Roll seems to be stocked with so many nutritionally desirable ingredients, I can't imagine why anyone would dub it "junk."

I also have difficulty with the idea that a baked potato smothered with butter and/or sour cream is an appropriate food, but slices of potato fried in vegetable oil and called "chips" become the work of the devil. In a similar vein, I find it difficult to believe that bread—I'm looking at a Pepperidge Farm white bread which contains as basic ingredients unbleached wheat flour, milk, water, vegetable shortening, salt, honey, and yeast—is a wholesome food, but leave out the yeast and the resulting crackers or pretzels become snacks and "junk" not fit for human consumption.

I suspect something else is at work here. Candy tastes good because one of its ingredients is sugar, and for some people sugar is the cause of tooth decay, diabetes, heart disease, and cancer, and totally unsuitable for growing children. Is this true? Is it based on supportable evidence? Let's take these charges one at a time.

Diabetes is not a product of modern society. It has been with us for a long time. Diabetes has been a medical problem for at least 2,000 years. It was described in the Ebers papyrus of ancient Egypt and was given its name by Aretaeus the Cappadocian in the first century A.D. In Asia, it was described by a Chinese physician in the year 200 A.D., and the Arab physician Avicenna was well aware of it. And today we are still baffled by it, although we know that it is a genetically linked metabolic disease in which the digestion of sugar occurs at a reduced rate or not at all because of a deficiency of a specific protein—insulin. In normal individuals, insulin is produced in sufficient quantity to metabolize all carbohydrates consumed. Heredity has long been recognized as an important factor in the occurrence of diabetes, and current evidence indicates that a faulty gene is the immediate cause of the condition. And the relationship obesity/diabetes also appears clear, but why this is so remains a mystery. One thing is certain: Although sugar, glucose, is improperly metabolized, ingestion of sugar does not produce the disease.

The belief that sugar is a cause of cancer of any body site or of any of the forms of heart disease is totally unsubstantiated.

The major forms of cancer—lung, colon, breast, prostate, uterus—are in no way related to sugar consumption. Eighty to 90 percent of lung cancer is clearly traceable to cigarette smoke. Another 10 to 20 percent is traceable to a variety of occupations and to radiation. Colon cancer may be related to lack of bulk in our diets, but as of this writing this remains to be established. Sugar consumption has never been even remotely implicated.

Breast cancer appears related to hormonal changes which in turn may be related to consumption of animal fat, but that too remains more of a hypothesis than an established fact. Sugar has never been implicated as a possible risk factor. And whatever the considered risk factors for other cancer sites—uterus, prostate, pancreas, etc.—sugar is not one of them.

Although sugar was suggested as a risk factor in heart disease by a British investigator in the early 1950s, it has since been discarded as a possible factor. Today, emphasis is on stress; even cholesterol, chased after for twenty-five years, is losing currency.

That leaves us with tooth decay—dental caries. Yes, it does appear that under certain circumstances, sugar promotes tooth decay in genetically susceptible individuals. But it is the sticky type of sugar-containing products, those that cling to teeth and gums, that are the culprits.

For this problem, two recommendations appear a lot more reasonable than banning "snacks." Two well-established preventives—fluoridation of water supplies or individual application of fluoride to the teeth, and simply brushing and rinsing the mouth with water—can offset the sugar-caries concern. In summary, it would appear that stigmatizing some foods as "junk" is based on little more than personal prejudices.

There is another phrase or idea that is making the rounds in the general fervor to do away with snacks and "junk" foods. That idea is "empty calories." "Empty calories" suggests that my Milky Way's 210 calories are without food value. Considering that most chocolate candy contains carbohydrate, proteins, and fat, the calories are hardly any more empty than carbohydrate, fat, and protein obtainable from chicken, fish, meat, or beans. This should not be construed to mean that candy or snacks should be substituted for poultry, fish, meat, or beans; it simply means that they, too, have a part to play in our overall nutrition.

Consider for a moment that we add celery, radishes, peppers (green and red), lettuce, cucumbers, and mushrooms to many salads, sandwiches, and side dishes. Why? Well, for color, for taste, for moisture, and to perk up meals. But what of the food value of these vegetables? There's hardly any. Other than contributing color—natural, of course—water up to 96 percent, some iron and calcium, there's precious little else in these foods. Talk of emptiness! Would anyone even suggest they be banned or removed? Hardly. They help make our meals tasty and they fill a need as quick snacks. Just like candy, potato chips, and soda. To coin a paraphrase, "junk" is in the mind of the abolisher.

I can only agree with Lewis Thomas, author of *The Snail and The Medusa* and chancellor of Memorial

Sloan Kettering Cancer Institute, who wrote, "The new danger to our well-being, if we continue to listen to all the talk, is in becoming a nation of healthy hypochondriacs, living gingerly, worrying ourselves half to death . . ."

6

Adulteration: The FDA's Eight Counts

Some years ago a fish dealer in Philadelphia wanted to get rid of—sell—a load of spoiled fish. To do so, he had to mask the "off" odor and color in order to make the fish appear "fresh." Accordingly, he liberally dosed the fish with sodium nitrate. Sodium nitrate, or Chile saltpeter, is one of the most abundant naturally occurring nitrates. When used properly, it is a rather innocuous substance. During the late Middle Ages it was discovered that the red color of meat became fixed when nitrate was added. It has been used for this purpose ever since. It is also well known that nitrates in sufficient quantity can cause a rapid drop in blood pressure through vasodilation—enlargement of the blood vessels.

In this instance, the fish was freely dosed with the nitrate and shipped out to markets along the Atlantic coast. Within a short time, 150 people became ill with nitrate poisoning; many were hospitalized and a three-year-old child died.

This is an unabashed example of adulteration—the debasing or lowering in quality of a food by adding inferior, cheaper, or less desirable materials to it. Interestingly enough, the dictionary says nothing about causing harm.

Sprinkling nitrate on fish to make it seem fresh when it patently is not is an example of one of eight forms of adulteration, specified by the Food and Drug Administration.

The FDA maintains that a food is adulterated if it contains any poisonous substance which may render it injurious to health. In such instances, the poison, whatever it is, must have been added—purposefully. Naturally occurring toxins or poisons, curiously enough, do not make the food adulterated under the law. That means that if peanut butter contains aflatoxin, a potent chemical produced by a fungus growing on peanuts, and those peanuts become peanut butter, and people eat that peanut butter and become ill—even die—that peanut buter cannot be called "adulterated." That's the law. If it's natural, it passes.

Before picking up on a second form of adulteration, a brief digression to explore the meaning of "poison."

Checking both standard and medical dictionaries, I find that "poison" is any substance dangerous to life and/or health. Here I anticipate quizzical expressions followed by the exclamation, "Well then!" or "By that, anything and everything can be poisonous if taken in sufficient quantity!"

Any number of examples pop into mind: aspirin, vitamins, penicillin, coffee (caffeine), alcohol, cigarettes, and certain proteins—even water. Of course, the Latins had a word for it—*"dosis sola facit venenum"*—it's the dose that makes the poison. The nitrate in the fish story is a grim example of this.

So, for something to be a poison, it has to be consumed in appropriate quantity, and this suggests that nothing is *inherently* poisonous but that everything can be. Now that *is* something to dwell upon.

For its second type of adulteration, the FDA tells us that if a food contains a food additive which is unsafe (as specified in section 409 of the Food and Drug Act) or if it contains a new animal drug or breakdown product of that drug, or a pesticide, that food may be adulterated. Why "may be"? Because section 409 also tells us that pesticide residues are not necessarily adulterations if the amount present conforms to a legally prescribed tolerance (there's that idea of dose again), or if the food additive conforms both in use and amount to its prescribed specifications.

What about the third and fourth forms of adulteration? These two go together, because the third says that a food may be considered to be adulterated if it consists in whole or part of any filthy, putrid, or decomposed material, or if it is unfit for food (the decision is based on what a reasonable person would consider putrid or filthy), while the fourth maintains that adulteration exists if the food has been prepared, packed, or even held under unsanitary conditions, during which time it can be contaminated to a degree that it becomes injurious to health. Quite obviously, the concept of adulteration will vary by person, place, and time. As we grow more sophisticated, we expect more. Considering that much of our food comes from animals, the soil, caves, oceans, and rivers, the burden and pressure to produce an ever cleaner product becomes greater too.

The fifth category is to be expected. It says that a food is adulterated if it is the product of a diseased animal or if the animal has died by means other than proper slaughter. The sixth deals with the container the food comes in. It stipulates that if the package or container is itself composed of any harmful or poisonous substance that may render it dangerous to health, the food is adulterated.

Because the use of radiation to preserve food has not been approved, any food intentionally subjected to irradiation is considered adulterated. That's number seven.

And to deal with omissions and substitutions, the eighth category of adulteration, section 402 insists that if a valuable constituent has been omitted in whole or in part, or if inferior substitutions have been made, or if damage or inferiority has been concealed, or if the contents have been packed to mislead by exaggerating volume or bulk, the food can be deemed adulterated. People have devoted so much creative energy to make things seem other than what they are.

Adulteration is not a product of our contemporary society. It has been going on since the first bag of lentils changed hands. By the fifteenth century, making food appear what it was not had reached an art form. Frontier America was a pushover for the pitchman and itinerant drummer. Many of our current laws and concerns are hangovers from those free-wheeling days.

Although adulteration is still practiced, it is fair to say that our food supply and production system is probably the most watched-over of any system in the country, and many knowledgeable people are convinced that ours is the most abundant, protected, and nutritious food supply in the world.

7

Prohibited Substances and the Interim List

Considering all the ingredients permitted in food as a consequence either of petitions attesting to their safety, or of their classification as GRAS, an impression could be obtained that no substance is ever barred from food. That would be a false impression. In fact, a number of substances are specifically prohibited and their use or inclusion would constitute adulteration of the product.

On the list of substances specified in the Code of Federal Regulations (CFR) as "Substances Prohibited from Use in Human Food" are:

> *Calamus and its derivatives.* The material obtained from calamus, the plant known as sweet flag *(Acorus),* was, as its name suggests, a flavoring agent.
>
> *Chlorofluorocarbon.* It was considered for use as a propellant in aerosol cans, but it, too, was unable to pass the toxicity test.
>
> *Cobalt and its salts.* Cobalt had been used in beer to prevent gushing and improve foam.

When it was determined that small amounts of cobalt could produce chest pains resembling heart attacks, cobalt was banned.

Coumarin. Found in tonka beans, where it imparted fragrance, it was used as a flavoring ingredient until it was found that it could disguise unpleasant odors.

Cyclamates. The sodium and calcium salts of cyclamic acid were developed as nonnutritive sweeteners. Studies on rats found them to be carcinogenic.

Diethyl pyrocarbonate (DEPC). DEPC was used to inhibit microbial growth and hence fermentation in both alcoholic and non-alcoholic beverages. It isn't any more.

Dulcin. With a sweetening capability some 250 times that of sucrose, dulcin was developed as another nonnutritive sweetener. Laboratory tests have not yet found it suitable for use in food.

Monochloracetic Acid. At one time this derivative of acetic acid was considered for use as a preservative. It no longer is.

Nordihydroguaiaretic Acid (NDGA). NDGA, found naturally in plants, was used for a time in fatty foods as an antioxidant to prevent rancidity.

P-4000. The designation "4000" stands for the fact that it has been judged 4,000 times as sweet as sugar (sucrose). It's another of the nonnutritive sweeteners that has not been approved.

Safrole. Because of its fragrance, safrole was considered for use in food. It is obtained from sassafras, and oil of sassafras contains some 75 to 80 percent safrole.

Thiourea. Also known as thiocarbamide, it was considered for use as an antimycotic agent—preventing growth of molds—but was found to produce tumors in rats.

In addition to ingredients that are totally banned, there are substances that are permitted in foods, but on an interim basis. This means that the FDA has not been totally satisfied with the information it has received or obtained and wants more data, and until it is satisfied, the material may be used but eyes are on it. Mannitol, saccharin, and brominated vegetable oil are in this category. Most recently, caffeine, which was on the GRAS list, was removed to the Interim List as a result of studies on rats. The implication here is that we want more data to reassure us about a substance's safety or lack of it—but we'll use it until additional data support a decision to ban it.

8

FDA's Classification of Label Ingredients: The Big Five

Any and all ingredients added to a food formulation can be classified as one of the following:

A. A food ingredient such as potatoes, meat, sugar, and milk.

B. Ingredients generally recognized as safe— GRAS. Literally hundreds of ingredients are in this category. For example, it includes ascorbic acid, lecithin, malic acid, sodium sorbate, potassium alginate, and zinc stearate.

C. Ingredients classified as certified colors. These are the synthetic colorants.

D. Ingredients in this group include the twenty-plus naturally occurring coloring agents.

E. Ingredients in this group include the regulated substances which have been approved by the FDA. This group includes such ingredients as kelp, polysorbate 60, xanthan gum, silicon dioxide, and hundreds of others.

Together, A, B, C, D, and E make up those things we call "food additives."

These tables that follow are examples of common foods and their spectrum of additives. The columns represent the additives as follows:

> A = food ingredient
> B = GRAS
> C = FDA-certified colors
> D = colors, exempt (natural)
> E = regulated additives

CAKE MIX
Betty Crocker Super Moist Devil's Food

	A	B	C	D	E
Sugar *	X				
BHA, BHT		X			
Baking soda		X			
Monocalcium phosphate		X			
Propylene glycol mono esters					X
Salt		X			
Mono- and diglycerides		X			
Dextrose	X				
Trisodium phosphate		X			
Calcium acetate		X			
Guar gum		X			
Cellulose gum					X
Egg white	X				
Artificial flavor					X
Modified corn starch					X
Wheat starch		X			
Isolated soy protein		X			

* Depending upon the petition, sugar can be classified as GRAS (B) or as a food (A).

CANNED SODA
Sprite

	A	B	C	D	E
Sugar	X				
Corn sweeteners		X			
Lemon/lime oil extract		X			
Citric acid		X			
Sodium citrate		X			
Sodium benzoate				X	

ABBOTTS CAKE COATED BAR (ICE CREAM)
Chocolate Eclair

	A	B	C	D	E
Milk Fat	X				
Nonmilk fat	X				
Corn sweetener		X			
Sugar		X			
Mono- and diglycerides		X			
Cellulose gum		X			
Locust bean gum		X			
Polysorbate					X
Guar gum		X			
Carageenan					X
Natural and artificial vanilla flavor		X			
Partially saturated soybean and coconut oils	X				
Lecithin		X			
Vanillin		X			
Salt		X			
TBHQ					X
Malt		X			
Caramel color				X	
Carob		X			
Bicarbonate of soda		X			
Niacin		X			
Vitamin B_1		X			
Vitamin B_2		X			
Iron (reduced)		X			
Coca	X				
Nonfat dry milk	X				

TANG

	A	B	C	D	E
Sugar	X				
Citric acid		X			
Calcium phosphate		X			
Malto-dextrin		X			
Potassium citrate		X			
Vitamin C		X			
Orange flavor		X			
Cellulose					X
Xanthin					X
Vitamin A palmitate		X			
BHA					X
Alpha-tocopheral		X			
FD&C Yellow #5			X		

GENERAL FOODS
INTERNATIONAL COFFEES
Swiss Mocha

	A	B	C	D	E
Sugar	X				
Hydrogenated coconut oil	X				
Corn syrup solids	X				
Instant coffee		X			
Cocoa (processed with alkali)		X			
Sodium caseinate solids		X			
Malto-dextrin		X			
Dipotassium phosphate	X				
Trisodium citrate		X			
Mono- and diglycerides		X			
Silicon dioxide				X	
Salt (common)		X			
Lecithin		X			
Tetrasodium pyrophosphate		X			

HERR'S
FRESH CRISP POTATO CHIPS

	A	B	C	D	E
Potatoes	X				
Vegetable oil	X				
One or more of the following:					
Cottonseed					
Soy					
Peanut					
Corn					
Safflower					
Salt			X		

HEALTH FOOD STORE ITEM
Tigers Milk

	A	B	C	D	E
Brewer's yeast *					
Tricalcium phosphate **					
Milk whey	X				
Sodium caseinate		X			
Nonfat dry milk solids	X				

---

* Although not listed by name in either CFR part 172, Regulated Food Additives, or Part 173, GRAS Additives, it may be the Dried Yeast noted in Part 172

** Under the Proxmire Amendment [+], the FDA is prohibited from looking at products called "nutrient supplements." Tricalcium phosphate, though not listed in either Part 172 or Part 182, Title 21 of the CFR, can thus be used in products such as Tigers Milk without fear of FDA control. Legally, it may be an adulterant.

[+] The Proxmire Bill amended the Food, Drug and Cosmetic Act, H.R. 7988, 94th Congress, 2nd Session, 1976.

HEALTH FOOD STORE ITEM
Whole Earth Center Bakery
Pecan Slice

	A	B	C	D	E
Butter	X				
Honey	X				
Whole wheat pastry flour *	X				
Eggs	X				
Pecans	X				
Molasses	X				
Vanilla		X			
Baking powder **		X			
Sea salt ***		X			

* Although Whole wheat flour is listed as a Standardized Food, the adjective "pastry" is not included in the legal definition.
** Calcium phosphate
    Potassium bicarbonate
*** A mixture of sodium chloride, magnesium chloride, magnesium sulfate, and calcium sulfate. This product has not been given a GRAS or regulated classification. Consequently, it is legally an adulterant. (See "Adulteration: The FDA's Eight Counts.")

HEALTH FOOD STORE ITEM
Fearn Soy/o
Pancake Mix

	A	B	C	D	E
Unbleached flour	X				
Soya powder*		X			
Potassium bicarbonate		X			

* Soya powder is not controlled by the FDA because "health foods" have been able to avoid FDA scrutiny. Legally it may be an adulterant.

HEALTH FOOD STORE ITEM
Whole Wheat Fig Bars

	A	B	C	D	E
Figs	X				
Whole wheat flour	X				
Corn syrup		X			
Kleen raw brown sugar	X				
Sugar	X				
Vegetable shortening		X			
Partially hydrogenated soybean oil	X				
Partially hydrogenated palm oil	X				
Partially hydrogenated cottonseed oil	X				
Dextrose	X				
Honey	X				
Salt		X			
Caramel color			X		
Whey		X			
Citric acid		X			
Glycerine		X			
Bicarbonate of soda		X			
Monocalcium acid phosphate		X			
Lecithin		X			
Ammonium carbonate		X			

HEALTH FOOD STORE ITEM
Hansen's Mandarin Lime Natural Soda
Bottled soda

	A	B	C	D	E
Fructose		X			
Citric Acid		X			
Mandarin flavor		X			
Lime flavor		X			

KELLOGG'S CORN FLAKES

	A	B	C	D	E
Milled corn	X				
Sugar	X				
Salt		X			
Malt flavor		X			
BHT					X
Vitamins					
C (sodium ascorbate)		X			
B (niacinamide)		X			
A (palmitate)		X			
B_6 (pyridoxine hydrochloride)		X			
B_1 (thiamin)		X			
B_2 (riboflavin)		X			
B (folic acid)		X			
D		X			
Iron (reduced)		X			

SUN COUNTRY GRANOLA

	A	B	C	D	E
Rolled oats	X				
Brown sugar	X				
Partially hydrogenated soya oil	X				
Almonds	X				
Corn syrup		X			
Soya flour	X				
Sweet dairy whey*		X			
Nonfat dry milk	X				

* In the Code of Federal Regulations, "whey" does not carry the descriptive modifier "sweet dairy." This is the manufacturer's preferred description.

PERCENT OF RDA'S

Sun Country Granola		Kellog's Corn Flakes
10	Protein	4
	Vitamins:	
*	A	25
2	B (niacin)	25
NI	B (folic Acid)	25
10	B_1 (thiamin)	25
6	B_2 (riboflavin)	25
4	B_6 (pyridoxine)	NI
*	C	25
NI	D	10
6	Calcium	*
10	Iron	10
20	Phosphorus	2
NI	Magnesium	*

* Provides less than 2%
NI—Not indicated

ARNOLD

Brick Oven White Bread

	A	B	C	D	E
Unbleached enriched spring * wheat flour	X				
Corn syrup		X			
Nonfat milk	X				
Salt		X			
Grade A creamery butter		X			
Golden Honey	X				
Calcium propionate		X			
Mono- and diglycerides		X			
Vitamins (niacin, thiamine, and riboflavin)		X			
Iron (reduced)		X			

* The "spring" in spring wheat is an optional description.

HEALTH FOOD STORE ITEM
Shiloh Farms
Sour Dough Caraway Rye Bread *

	A	B	C	D	E
Stone ground ** whole wheat flour	X				
Rye meal	X				
Stone ground rye flour	X				
Salt		X			
Yeast		X			
Caraway		X			
Malt		X			

* Sold as a frozen loaf.
** "Stone ground" is an optional description with no legal status.

PERCENT OF RDA'S *
PROVIDED BY EACH LOAF

Arnold		Shiloh Farms
2	Protein	5
2	Iron	6
2	Calcium	2
	Vitamins:	
0	A	0
4	B (niacin)	4
6	B_1 (thiamin)	8
4	B_2 (riboflavin)	4
0	C	0

* Recommended Daily Allowances

9

Types of Additives

Acidulants are used to impart tartness or maintain required acidity or alkalinity—pH control.

> Acetic acid
> Adipic acid
> Citric acid
> Fumaric acid
> Lactic acid
> Malic acid
> Phosphoric acid
> Sodium acetate
> Sodium bicarbonate
> Sodium citrate
> Tartaric acid

Anticaking substances are used to keep seasoning salts and a wide variety of powdered mixes from turning into solid chunks in damp weather.

> Calcium stearate
> Calcium silicate
> Iron ammonium citrate
> Magnesium carbonate
> Mannitol
> Silicon dioxide
> Sodium alumino silicate
> Tricalcium phosphate
> Yellow prussate of soda

Antioxidants are ingredients used to delay or prevent fats and oils from becoming rancid, or to prevent or delay cut fruits and vegetables from discoloring—turning brown, black, or gray.

> Ascorbic acid—Vitamin C
> BHA—Butylated hydroxy anisole
> BHT—Butylated hydydroxy toluene
> Propyl gallate
> TBHQ—Tertiarybutylhydroquinone
> THBP—Trihydroxybutyrophenone
> Tocopherols—Vitamin E

Dough Conditioners are ingredients that modify proteins and cellulose by reducing the 'toughness' or 'springiness' of dough, making it easier to handle and more palatable to eat.

> Ammonium monocalcium phosphate
> Azodicarbamide
> Benzoyl peroxide
> Calcium stearyl-2-lactylate
> Calcium sulfate
> Glyceryl monostearate
> Hydrogen peroxide
> Potassium bromate
> Sodium stearyl fumarate
> Sodium stearyl-2-lactylate

Emulsifiers are used to prepare mixtures of oil and water that will mix, and stay mixed.

> Arabinogalactose
> Dioctyl sodium sulfosuccinate
> Lecithin
> Mono- and diglycerides
> Polysorbate 20, 40, 60, 65, 80

Propylene monostearate
Sodium lauryl sulphate
Sodium silicate
Sorbitan monostearate

Flavoring and Flavor-Enhancing ingredients are used to increase flavor.

Bacterial starters
Disodium guanylate
Disodium inosinate
HVP—Hydrolyzed Vegetable Protein
Hop extract
MSG—Monosodium glutamate
Paprika
Quinine
Safrole—free extract of Sassafras
Sugar beet extract
Turmeric
Vanillin
Yeast-malt sprout extract

Humectants help retain texture by preventing and/or retarding loss of moisture. The name comes from a Latin verb meaning *to moisten.*

Glycerine
Glycerine monostearate
Mannitol
Propylene glycol
Sodium tripolyphosphate
Sorbitol

Leavening Agents are added to increase volume and thus to produce light and fluffy baked products.

> Ammonium carbonate
> Calcium phosphate
> Sodium aluminum phosphate
> Sodium aluminum sulphate
> Sodium bicarbonate
> Sodium silico aluminate

Nonnutritive Sweeteners replace sugar or corn syrup in dietetic foods. Calories are not provided.

> Aspartame
> Sodium saccharin
> Cyclamate

Nutrient Supplements are added to increase food value in naturally deficient foods as well as those that lose nutrients during processing.

> Aluminum nicotinate
> Amino acids
> Bakers' yeast protein
> Calcium pantothenate
> Folicin
> Iron choline citrate complex
> Kelp
> Mannitol
> Potassium iodide
> Vitamins
> Whole fish protein concentrate
> Xylitol

Nutritive Sweeteners are used for taste enhancement as well as their food value. They provide calories—food energy.

> Corn syrup
> Dextrose
> Fructose
> Glucose
> High-fructose corn syrup
> Honey
> Maltose
> Maple syrup
> Sucrose

Preservatives are used to prevent or retard microbial (bacterial, fungal, and yeast) spoilage.

> Ascorbic acid
> Ascorbyl palmitate
> Benzoic acid
> Calcium propionate
> Calcium sorbate
> Paraben (methyl, ethyl, butyl, heptyl)
> Potassium sorbate
> Propionic acid
> Sodium benzoate
> Sorbic acid
> Sodium diacetate
> Sodium nitrate
> Sodium nitrite

Sequestrants combine with traces of metals that can produce off-odors and off-tastes. Sequestering (sequestrant) comes from the verb *sequester,* which means to remove, withdraw, or separate. In fact, they withdraw metals from a variety of foods.

  Calcium diacetate
  Citric acid
  EDTA (Ethylenediaminetetraacetic acid)
  Phosphoric acid
  Sodium metaphosphate

Stabilizers/Thickeners enhance viscosity and "mouth-feel," and prevent emulsions from separating.

  Agar
  Arabinogalactan
  Calcium alginate
  Carageenan
  Carob bean gum
  Disodium phosphate
  Guar gum
  Gum acacia
  Gum arabic
  Gum ghatti
  Gum tragacanth
  Karaya gum
  Locust bean gum
  Propylene glycol
  Propylene glycol alginate
  Sodium alginate
  Sodium methyl cellulose
  Xanthan gum

10

Additives from A to Z

A

Acetic Acid and vinegar are similar, but they are not the same. The acidic ingredient of vinegar is acetic acid, which in pure form is used in catsup, mayonnaise, and pickle products for flavor and antimicrobial activity. Has been used in foods since 300 B.C. *GRAS*

Acetone Peroxide is used to bleach and "mature" flour after milling. (See Benzoyl Peroxide.) *Regulated*

Adipic Acid is an acidulant that imparts a smooth, tart taste. Although it is found naturally in beet juice, much of what is used is manufactured, primarily for use in dry fruit drink powders and gelatins. It is also used in meats and sausages as a preservative. *Regulated*

Aluminum Nicotinate is a nutritional supplement added to special dietary foods to increase the available amount of niacin. To avoid being mistaken for nicotine, it should be called niacinamide. *Regulated*

Amino Acids Of the eighteen amino acids contained in food proteins, eight are essential for health and must be contained in our diets, as the body is incapable of synthesizing them. In this group are tryptophan, phenylalanine, lysine, threonine, methionine, leucine, isoleucine, and valine. Two, histidine and arginine, are semiessential in that they are synthesized but in inadequate amounts, and six others are nonessential, as they can be synthesized by the body.
Regulated

Ammonium Carbonate, although a general-purpose additive (which means it can be added to food to achieve a number of qualities) is none other than the 'spirit of hartshorn' that Grandma used as a leavening ingredient in her cakes.
GRAS

Ammonium Monocalcium Phosphate is a general-purpose additive. It can be used to affect acidity and moisture retention and as a dietary supplement.
GRAS

Anise is a flavoring ingredient obtained from the plant *Pimpinella anisum.* Licorice is the best way to describe its taste.
GRAS

Annatto, or Bixin, is an extract obtained for its color from a tropical tree *Bixa orellana.* Annatto's color ranges from butter-yellow to peach.
GRAS

Arabinogalactose (Galactan) is incorporated into foods for its emulsifying, binding, and/or bulking qualities—particularly in puddings. This is a complex carbohydrate, a polysaccharide originally obtained from the larch, a member of the pine family.
Cleared

Ascorbic Acid, also known as vitamin C, the antiscor-

butic ingredient, is added to food for its ability to speed up color-fixing in cured meats, as well as its ability as a preservative.
GRAS

Ascorbyl Palmitate, a derivative of ascorbic acid (vitamin C), is added to fatty foods to retard rancidity.
GRAS

Aspartame, or aspartyl phenylalanine, was constructed by linking two amino acids which occur naturally in both plants and animals. Alone, neither is sweet, but the compound appears to be 200 times sweeter than sucrose. On July 15, 1981, Arthur Hayes, Jr., M.D., Commissioner of the FDA announced that Aspartame would be cleared for use as a non-nutritive sweetener, *Nutra-Sweet*. It has been cleared for use in cold cereals, drink mixes, instant coffee and tea, gelatin, pudding fillings, dairy products and dessert toppings. Aspartame will probably become available late in 1981 or early 1982.
Cleared

Azodicarbamide is a relative newcomer for use in dough maturing. It acts quickly and is often used in combination with potassium bromate.
Cleared

B

Bacterial Starters are harmless cultures of bacteria added to pork rolls, salami, Thuringer sausage, Lebanon Bologna, and cervelat to develop additional flavor.
Cleared

Bakers' Yeast Protein is obtained from yeast and used in protein deficient food as a nutritive supplement.
Cleared

Benzoic Acid is an organic acid found naturally in cranberries, prunes, plums, and cinnamon. It is used as both a flavoring and an antimicrobial preservative.
GRAS

Benzoyl Peroxide is used to "mature" or age flour by modifying (oxidizing) proteins that lead to better handling characteristics and larger loaf volume. Freshly milled flour is yellow and has poor baking qualities. Benzoyl peroxide is used to speed up the otherwise long storage periods necessary to produce white flour and optimum baking properties.
Cleared

Beta-Apo-8-Carotenal sounds as though it ought to do more than color foods. Actually, it is one of the most widely available natural substances. It is one of the carotenoids, which are responsible for the color in carrots, apricots, lobsters, and orange juice.
Cleared

BHA—Butylated Hydroxy Anisole, also known as Embanox, is an antioxidant used to prevent or retard the rancidity that can occur when oxygen and air combine with oil and fats.
Cleared

BHT—Butylated Hydroxy Toluene, first prepared in 1949 as a waxy solid, is noted for its antioxidant properties. It has a synergistic effect with acids, lecithin and other substances including BHA that increases its effectiveness. At a recent meeting, a joint committee of the World Health Organization and the FAO (Food and Agriculture Organization of the UN) agreed to remove the temporary ADI—allowable daily intake—for a permanent one on the basis of new data showing it to be safe.
Cleared

C

Cajeput is a pungent oil obtained from the East Indian "paper bark" tree—*Melaleuca leucadendron*—used to flavor a variety of foods.
GRAS

Calcium Alginate is one of the gelling substances—gums—obtained from the giant kelp *Macrocytis pyrifera*, harvested along the California coast. The alginates are used for their water-binding, gel-forming, and emulsion-stabilizing power. They are particularly stable in acid foods such as salad dressings. In cheeses, they reduce surface hardening.
GRAS

Calcium Chloride, discovered in the fifteenth century, wasn't used in food until the twentieth. It was first used to prevent canned vegetables from excessive softening by cooking. It is common practice to add calcium chloride to tomatoes, apples, and other vegetables prior to canning or freezing. Like its sodium cousin, this too is inorganic.
GRAS

Calcium Diacetae (See Disodium EDTA.)

Calcium Ethylenediaminetetraacetate (EDTA) (See Disodium EDTA.)
Cleared

Calcium Gluconate (See Sodium EDTA.)
GRAS

Calcium Pantothenate is a derivative of pantothenic acid, one of the B vitamins. It is added to special dietary foods as a nutritional supplement.
GRAS

Calcium Phosphate, sometimes referred to as acid calcium phosphate and monocalcium phosphate, is a general-purpose additive that is used at times for its sequestering ability and at other times as a dietary supplement. This is a good example of the use of an additive for multiple effects, depending upon the type and amount of other ingredients with which it is mixed. As a sequestering agent, it binds trace metals that could cause off- odors and tastes.

GRAS

Calcium Propionate is an acidulant preservative with fungistatic properties. This is the calcium salt of propionic acid, a weak organic acid—one of the "goat acids," so-called because of their strong odor. In dilute concentrations it has a slight cheeselike odor. It is used to protect processed cheeses and baked goods from mold spoilage.

Regulated

Calcium Silicate is another anticaking ingredient used in baking powders, dry mixes, and table salt to maintain their free-flowing properties.

GRAS

Calcium Sorbate is one of the salts of sorbic acid. Although it is a "salt," it does not contain sodium. Because it has broad antimicrobial activity against yeasts and molds, it is used in a wide variety of foods—cheeses, pickles, beverages, and baked products. It is obtained from the berries of the mountain ash.

GRAS

Calcium Stearate has anticaking properties. In seasoning salts and other powders that absorb moisture from the air, it prevents their turning into a solid chunk. It is derived from edible tallow.

Regulated

Calcium Stearyl-2-Lactylate is a derivative of lactic acid. The FDA has approved it as an optional ingredient for maturing bread dough. It is also used to

enhance the whipping quality of toppings.
GRAS

Calcium Sulfate is a multipurpose ingredient. In doughs it serves as food for yeast to stimulate gas production; it also aids in the rapid maturing of flour, as a firming agent in certain canned vegetables, and as a dietary supplement to increase available calcium.
Cleared

Candelilla Wax is obtained from the Euphorba, a cactuslike plant. It is used as a flavoring and coating in candies and confections—it has the ability to remain solid for extended periods at body temperature. It is the ingredient that keeps candy from melting in your hand.
GRAS

Canthaxanthin is one of the carotenes, a naturally occurring orange-red coloring agent. The Commissioner of Food and Drugs has concluded from available test data that there is no basis for concern about this or any of the others listed under Natural Coloring Ingredients (page 30). As a consequence of its high tinctorial character, it is used in tomato products, barbecue sauces, spaghetti sauce, cheeses, and shrimp, salmon and lobster products, to note a sampling.
Cleared

Carob Bean Gum (See Locust Bean Gum.)
GRAS

Carminic Acid, known also as cochineal extract, is a red coloring ingredient from the dried bodies of an insect (female only), *Coccus cacti,* found primarily on cactus in the Canary islands. (See Coloring Ingredients, Natural Products, page 30.)
Cleared

Carrageenan is obtained from marine algae called Irish moss that grow in tidal pools along rocky seacoasts. The United States, France, and Denmark are major producers. Carrageenan is used for ice-cream stabilization in dairy products; for suspension of cocoa powder in chocolate milk; in flans (milk-based starch puddings), whipped toppings, and coffee whiteners; and to obtain proper "mouth feel" in frozen fruit concentrates and fruit drink powders.
Cleared

Citric Acid has been used to achieve tartness in foods for over 100 years. It is also used to help retard rancidity. The FDA classifies citric acid as a general-purpose food additive.
GRAS

Cyclamate, Calcium is a nonnutritive, noncaloric synthetic sweetener used in place of sugar. Cyclamate, calcium, and/or sodium, has been banned by the FDA. On September 16, 1980, Abbott Laboratories of Chicago indicated it would no longer continue its struggle to have cyclamate recertified. At the present time, Canada, West Germany, Sweden, Norway, and Switzerland permit the use of cyclamates in food.
Banned

D

Dextrose (See Glucose.)
GRAS

Dicalcium Phosphate, an inorganic compound, is a leavening acid * used to produce gas late in the baking cycle. It has limited use and is consequently used little.
GRAS

* See Baking Soda

Diglyceride (See Glyceride.) Over 98 percent of fat naturally present in food is in the form of glycerides. Most fats contain at least two different fatty acids (FAs) and are therefore *mixed glycerides.* The remaining 2 percent of food fat consists of *mono-* and *diglycerides.* The monos are primarily glycerol and one FA, and the diglycerides are glycerol and two FAs. All function as *emulsifiers*—keeping fat in finely divided form. All added mono- and diglycerides are chemically similar and function in the same way.
Cleared

Dimethyl Polysiloxane (DMPS) During the manufacturing process, foaming or frothing can occur in some foods. DMPS is used as an antifoaming ingredient in dairy products, soups, starches, and baked goods.
Regulated

Dioctyl Sodium Sulfosuccinate DSS is a surfactant that permits rapid wetting of dry ingredients as well as better whipping qualities in toppings. It increases foaminess where wanted and can decrease it when not wanted.
Cleared

Disodium Ethylenediaminotetraacetate (EDTA) is one of the most widely used *sequestrants,* substances that react with trace metals (naturally present in food ingredients) to form complexes that prevent the metal from entering into chemical reactions, i.e., cobalt in vitamin B_{12}, magnesium in chlorophyll, iron in hemoglobin. Sequestrants are used to stabilize fats and oils which undergo rancidity and reversion in the presence of the metals copper and iron. By chelating (sequestering) metals, oxidation is slowed or entirely prevented. Vitamins are notorious for their instability and loss of potency. EDTA is able to prevent vitamin breakdown—particularly the fatty vitamins A, D, E, and K. Other chelating agents used are:

> Calcium diacetate
> Calcium gluconate
> Sodium tartrate
> Sodium acid pyrophosphate
> Sodium hexameta phosphate
> Tetra sodium pyrophosphate

Disodium Guanylate and Disodium Inosinate both are flavor potentiators. These compounds are known as nucleotides and come from the same family as DNA and RNA. They are used in dry soup mixes, cereals, and meat and fish products to enhance flavor. Although potentiators enhance flavor they are not the same as enhancers such as MSG in that they are much more powerful and thus are used in far smaller quantities. And again, it was a Japanese, Dr. Shintara Kodama, who discovered these in the early years of this century.
Cleared

Disodium Phosphate has mildly alkaline properties. It is used in foods to buffer or prevent shifts to either a more acidic or a more basic condition.
Cleared

E

Erythorbic Acid also known as isoascorbic acid is used as both a preservative and a color-fixing ingredient in cured and comminuted meats.
GRAS

Ethoxyquin is an antioxidant used for the preservation of color in the production of chili powder, paprika, as well as in ground chili.
Cleared

F

Ferrous Gluconate is a "salt" (no sodium in this salt) of gluconic acid (a derivative of glucose) used for its ability to contribute iron to foods lacking it in adequate amounts. By attaching the gluconate the ordinarily insoluble iron becomes available to the body.
GRAS

Fluorinated (and Chlorinated) Hydrocarbons are gases used to push or propel whipped-cream topping from cans. They also produce a fluffiness, exclude oxygen, and prolong shelf life.
Regulated

Folacin, another name for folic acid, is one of the B vitamins. Chemically, it is called Pteroryl glutamic acid and is used to enhance the nutrient content of deficient foods.
Cleared

Fructose, also called levulose or fruit sugar, is the sweetest of all natural sugars. It is 1½ times sweeter than sucrose, the standard of sweetness. Although it is a naturally occurring substance, it is still a food additive.
GRAS

Fumaric Acid is an acidulant that is used in a wide variety of foods such as lard, butter, cheese, powdered milk, bacon, franks, nuts, and potato chips for its rancidity-retarding ability. It takes its name from the plant *Fumaria officinalis*.
Cleared

Furcellaran is obtained from a marine plant *Furcellaria fastigiata*. Is used as an emulsifier and thickener in blancmange puddings, and as a stabilizer to reduce "weeping" in milk puddings.
Cleared

G

Glucono Delta Lactone is used primarily as a leavening agent in yeast-type instant bread. It is also used in doughnuts to reduce amount of grease taken up.
Cleared

Glucose, also called dextrose, is the primary form into which sugars are converted in the body. Thus, it is the principal sugar found in blood. Glucose is naturally present in many fruits.
GRAS

Glycerides (Mono- and Di-)are fatty ingredients used in a wide variety of foods for their ability to produce smooth texture. Because they are soluble in both water and oil, they also make excellent emulsifiers. They also have the ability to keep foods moist.
Cleared

Glycerol Esters of Rosin An "ester" is nothing more than a unique arrangement of organic molecules. Glycerol esters are used primarily in chewing gum to give the resistance necessary for proper "mouth feel" when chewing. The "ol" of glycerol indicates that it is naturally sweet but it is not sugar. All the sweet polyols have the ability to absorb water and become viscous. Rosin, known for centuries, is obtained from the exudate of pine trees.
Regulated

Glyceryl Monostearate (See Diglyceride for an explanation of this *monoglyceride.*)
GRAS

Guar, obtained from the seed of a legume, resembles the soybean, which is grown widely in India and Pakistan and now the United States. It has the ability to take up (hydrate) cold water quickly and attain a high viscosity (thickness). Thus it is used primarily for its water-binding ability. Guar is used as a stabi-

lizer in ice cream, doughs, and baked goods, and to control thickening in beverages, salad dressings, and relishes.
GRAS

Gums Four groups of (gums) can be used to classify all those used in foods:

Vegetable gums—guar, locust bean, tamarind seed, tragacanth

Marine seaweed gums—carrageenan, alginate, agar

Microbial fermentations—xanthin

Synthetic—methyl cellulose

Gum Acacia is actually the dried exudate of the acacia tree native to the Middle East, and has been used by the Egyptians as a thickener in foods for over 3,500 years. Currently it is used primarily for its ability to stabilize foams and ice cream.
GRAS

Gum Arabic (See Gum Acacia.)
GRAS

Gum Ghatti is obtained from large trees in India when the bark is damaged. It is used primarily as an emulsifier and also when high viscosity is desired.
GRAS

Gum Tragacanth is another gum obtained from trees in the Middle East. It too has been used in foods for hundreds of years. It is used primarily as a stabilizer and thickener for salad dressings, ice cream, sauces, and candies.
GRAS

H

High Fructose Corn Syrup (HFCS) is a sweetener made from corn starch in which the fructose content has been increased. Since fructose is the sweetest sugar, HFCS is sweeter than regular corn syrup.

Hop Extract is a natural flavor ingredient obtained from the plant *Humulus lupulus,* used in the brewing of beer to impart beer's characteristic bitter taste and pleasant aroma.
GRAS

Hydrogen Peroxide is a common oxidizing agent. It is usually limited in its use to whiten the color of milk for cheese manufacture or to bleach tripe.
GRAS

Hydrolyzed Vegetable Protein (HVP) refers to a group of amino acids obtained by the hydrolysis or splitting of a variety of plant or vegetable proteins. HVP is used as a flavor enhancer in soy sauce, for example.
GRAS

I

Invert Sugar is simply a mixture of two sugars, glucose (dextrose) and fructose (levulose), resulting from the splitting of sucrose. The importance of invert sugar in confections (caramels) is its ability to prevent crystallization of sucrose. Substituting part of the sucrose with invert sugar reduces the likelihood of crystallization, as both glucose and fructose crystallize more slowly than sucrose.

The term *invert* refers to the fact that when a solution of sucrose changes to a mixture of fructose and glucose, there is an accompanying change in the rotation of light from left (-) to right (+).
GRAS

Iron Ammonium Citrate, also known as green ferric ammonium citrate, is an anticaking ingredient used in salt to keep it free-flowing in periods of high humidity.
Cleared

Iron-Choline Citrate Complex, for all its lengthy name,

is simply another means of adding iron to foods deficient in this important mineral.
Cleared

K

Karaya Gum was first obtained from an Indian tree *Sterculia urens*. The gum is a complex carbohydrate and is used in food as a stabilizer for salad dressings, sherbets, and whipped-cream items. It can also be used as a binder in meat products.
GRAS

Kelp is the gum obtained from a brown marine alga *Laminaria digitata,* found along rocky Atlantic coasts. Along with its ability to contribute bulk to food, its high iodine content is used to provide protection against hypothyroidism—goiter.
GRAS

L

Lactic Acid is one of the most widely distributed acids in the plant kingdom. It is used in a host of foods to adjust acidity.
GRAS

Lecithin, also known as phosphotidyl choline, is a mixture of the diglycerides of stearic, palmitic and oleic acids linked to phosphoric acid. It is one of the most common phosphorous-containing fatty acids found in nature. Egg yolk, for example, can contain as much as 10 percent lecithin. Soybean lecithin is used in foods as both an emulsifier and an antioxidant.
GRAS

Locust Bean Gum is also known as carob bean gum and St. Johns Bread. It is used as a stabilizer in sherbets, soft cheeses, and whipped toppings.
GRAS

M

Magnesium Carbonate is one of those hygroscopic substances that have the ability to absorb quantities of moisture and keep other substances dry. In this case, magnesium carbonate, a naturally occurring "salt" obtained from the minerals magnesite and calomite, is added to powders to keep them from caking.
GRAS

Magnesium Stearate is one of the "release" agents used in foods to prevent baked products and candies from sticking to themselves and containers. Actually it is a fatty or oily powder.
GRAS

Malic Acid is found naturally in many fruits and vegetables, but is also made commercially. It is one of the general-purpose acidulants. It has been used in foods for decades.
GRAS

Maltose, or malt sugar, is a product of the fermentation of starch. Barley malt, used in brewing, enhances the flavor and color of beer because of its maltose content.
GRAS

Mannitol is a polyhydric alcohol, a *polyol*. Because of its ability to absorb and retain water under conditions of medium relative humidity, it is used to keep foods moist, and imparts needed bulk. Derives its name from the sweet manna of the biblical story.
Regulated

Methyl Glucoside—Coconut Oil Ester is used for its dual ability as a surface-active agent in molasses and as an aid in the crystallization of sugars. Coconut oil is highly saturated, making it highly desirable for use in shortenings, margarines, cake mixes, and pressurized toppings.
Cleared

Monocalcium Phosphate is a leavening acid used for the slow evolution of gas in pizza doughs, pancake mixes, and angel food cakes.
GRAS

Monoglyceride (See Diglyceride.)
Cleared

Monosodium Glutamate, an amino acid, is probably the best known of the flavor enhancers, substances that intensify the flavor of other ingredients without imparting any of its own. Its flavor-enhancing properties were discovered at the turn of the century by Dr. Ikeda at the University of Tokyo, as a consequence of a search for the flavor-enhancing properties of the seaweed *Laminaria,* used in Japan for centuries to improve the flavor of soups and other foods. In May, 1980, a scientific review committee set up by the FDA reported that MSG presented no hazard to the health of adults at present levels of use. However, the committee maintained that manufacturers should show restraint in adding MSG to food because people have reported reactions to it, even though the reactions are in fact harmless. The reactions in question are those grouped together as the "Chinese Restaurant Syndrome"—headache, burning sensations on neck and forearm, tightness in chest and neck, and pressure behind the eyes. Different people report one or two of these symptoms. Because of the process used to make MSG, it contains a good deal of sodium and tastes salty.
GRAS

O

Oxystearin is one of the few ingredients used in vegetable oils to inhibit crystal formation. Oil-water emulsions can form a haze or cloudiness if refrigeration

temperatures are too low. Oxystearin prevents clouding.
Cleared

P

Paprika Oleoresin is the fat-soluble coloring material extracted from paprika—a type of sweet pepper—for use in coloring food shades of red. Primarily used in salad dressings.
GRAS

Paraben is a family of compounds derived from p-hydroxybenzoic acid and as such it has properties similar to benzoic acid. In this group are methyl, ethyl, propyl, and butyl paraben. These substances have antimicrobial activity and are therefore used to prolong the keeping quality of cakes, piecrusts, icings, toppings, and fruit fillings.
GRAS

Pectin is a water-soluble polysaccharide (a complex carbohydrate) found in land plants. In foods, it is used as an aid in forming gels.
GRAS

Peroxidase is a protein enzyme used to remove glucose from dried egg products in order to increase their storage life.
Cleared

Phosphoric Acid is a general-purpose additive. Its primary function is as an acid in soft drinks to enhance flavor. "Phosphate," the southern and western name for soda, got its name from phosphoric acid.
Regulated.

Polysorbate-60 (20, 40, 65, 80) is a surface-active agent, or *surfactant,* that reduces "tension" at oil/water surfaces. This reduction of tension permits oils and water to mix. This mixing is called emulsification.

Polysorbate-60 also prevents fats from smoking and spattering when frying. It is used in peanut butter to prevent separation of the high oil content from peanut fiber. This is the additive that maintains mixtures of oil and vinegar in salad dressings long after you've stopped shaking the bottle.
Regulated

Polyvinyl Pyrrolidone is used on fruits, candies, cakes, and cookies for protection and appearance. It creates a bright, shiny surface and is a clarifying agent in beverages.
Regulated

Potassium Acid Tartrate (Cream of Tartar) is one of the first acid ingredients used in baking powders. It was first marketed in 1850.
GRAS

Potassium Bromate As a conditioner in doughs, it reduces toughness and springiness by modifying both the protein and cellulose in the dough. This modification makes the dough easier to handle as well as increasing its taste and "mouth feel."
Cleared

Potassium Gibberellate is simply a nutrient for the yeasts that are responsible for rapid and complete fermentation of doughs.
Cleared

Potassium Iodide is used primarily as a supplement to increase the level of iodine in foods—a preventive against goiter.
GRAS

Potassium Nitrate, an inorganic compound, is used as a curing ingredient in the processing of cod roe.
Regulated

Potassium Sorbate is a derivative of sorbic acid. It is

used to prevent spoilage by bacteria and yeasts. (See Sorbic Acid.)
GRAS

Propionic Acid is a fatty acid found naturally in Swiss cheese. It is used widely in bread doughs to inhibit mold spoilage.
GRAS

Propylene Glycol is used for its emulsifying ability in ice cream and for its ability to inhibit formation of crystals in icings and toppings.
GRAS

Propylene Glycol Monostearate has several functions. Although primarily a surface-active agent, it also contributes to tenderness in baked goods by its ability to trap air in batter mixtures to improve volume and texture. It also keeps fats and oils from separating.
Cleared

Propyl Gallate is an antioxidant used to retard rancidity. It is often used in combination with BHA and/or BHT because of its synergistic effects.**

Meeting in Rome during March and April of 1980, an Expert Committee on Food Additives of the World Health Organization and the Food and Agriculture Organization of the UN agreed to delete the "temporary" status of Propyl Gallate for a permanent ADI—allowable daily intake—as adequate data on its safety had been obtained.
GRAS

Q

Quinine, obtained from the bark of the cinchona tree, is used as a flavoring agent in carbonated beverages.
Cleared

s

Saccharin is a nonnutritive sweetener. It was discovered and synthesized at Johns Hopkins University in 1879. It is 500 times as sweet as sugar—which means a lot less is needed to achieve a similar level of sweetness. After almost a hundred years of use, a controversy currently exists as to its safety for human consumption. A ban on its further use is being considered by the FDA. Although it has been called a "low-risk" ingredient as a result of animal tests, the FDA believes it may be a potential human carcinogen. It has been placed on the FDA's Interim List for further study.

The moratorium on the bill to ban saccharin expired on June 30, 1981. Indications are that bills introduced to extend the moratorium will pass.
GRAS

Saffron is a natural coloring ingredient obtained from the dried and powdered stigmas of the perennial *Crocus sativus*. It should not be confused with meadow saffron (safflower) or bastard saffron. Saffron is one of the world's oldest known and most expensive spices.
GRAS

Silicon Dioxide is an inorganic compound (no carbon, no hydrogen) added to food for its ability to keep powders dry during moist weather; it prevents caking. It also has the ability to stabilize beer—to keep beer solids from settling out.
Cleared

Sodium Acetate is a derivative of acetic acid, also used as an acidulant for its tartness.
Cleared

Sodium Acid Pyrophosphate (SAPP) Although classed as a slow-acting leavening agent, it actually releases

60 to 70 percent of its gas (carbon dioxide) quickly. SAPP is replacing SAS in some baking powder formulations because SAS contributes unwanted flavor and can increase rate of rancidity.
Cleared

Sodium Aluminum Phosphate (SALP) is a widely used baking acid gaining the dominant position among leavening agents. It increases tenderness as well as firmness of crumb.
GRAS

Sodium Aluminum Sulphate (SAS) is an ingredient in double-acting baking powders. It is used to suppress gas formation in doughs until oven temperatures are reached. This way gas production is even and complete. Because it can increase the rate of rancidity, it is being phased out and replaced with SAPP.
Cleared

Sodium Benzoate has long been known for its anti-microbial properties. The sodium makes benzoic acid totally water-soluble and able to exert its anti-bacterial activity in food and soft drinks.
Cleared

Sodium Bicarbonate, baking soda, is common to all leavening formulations. The combination of "baking soda" ($NaHCO_3$) with an acid produces the gas needed for expanding dough.
GRAS

Sodium Carboxymethylcellulose is derived from cotton linters. It is used in jellies, pie fillings, sherbets, ice pops, and frozen confections to prevent growth of ice crystals, and in cakes to retain moisture.
Cleared

Sodium Caseinate is the protein of milk, found primarily in whey, and in the "skim" when milk is heated. Used to fortify and give texture to high-pro-

tein diets, nondairy coffee creamers, whipped toppings, imitation meat loaves, stews, and soups.
Cleared

Sodium Citrate or trisodium citrate is a white, odor-
less powder used for its emulsifying ability in pas-
teurized cheese and cheese spreads, to prevent
separation of fat and water as well as to achieve
smoothness of taste.
GRAS

Sodium Diacetate is often found in bakery products,
where it is used to prevent bacterial spoilage. It is a
derivative of acetic acid. Before the addition of sodium diacetate, "ropy" and stringy dough occurred
regularly as a consequence of bacterial growth.
GRAS

Sodium Hexametaphosphate (See Calcium EDTA.)
GRAS

Sodium Lauryl Sulphate is used in a wide variety of
foods for its emulsifying and surface tension–reducing capabilities.
Regulated

Sodium Metaphosphate, the more common name of
one of the polyphosphates, has been available for
over 150 years. It is used for its water-binding mois-
ture-retentive ability. It is also able to act as a sequestering agent by which it binds trace metals that
could produce off-odors and -colors. And it is also
used in freezing of poultry to prevent dripping during thawing.
GRAS

Sodium Nitrate (Saltpeter) is a natural ingredient of
our biosphere, found in soils around the world as a
water-soluble inorganic salt. It has been used to cure
meats, fish, and poultry—to fix their color and prevent bacterial growth—since late Roman times. By

the 13th century, sodium nitrate was in common use for curing meat. When added to meats, the nitrate is converted to nitrite by bacterial metabolism. It is the nitrite that keeps the meat looking its natural red.
Regulated

Sodium Nitrite Adding nitrite directly to meats, rather than the nitrate, which must be converted, assures greater control of quality. However, it was learned that nitrites could combine with amines to form potentially carcinogenic nitrosoamines. The controversy surrounding nitrite was calmed when on August 19, 1980, Dr. Jere E. Goyan, Commissioner of the FDA, and Carol Tucker Foreman, Assistant Secretary of Agriculture issued a Press Release saying there was "no basis for the FDA or the USDA to initiate any action to remove nitrite from foods at this time."
Regulated

Sodium Silicate is not added directly to food. It migrates to food from cotton fabric used in dry food packaging.
GRAS

Sodium Silico Aluminate (SSA) is really responsible for that great line, "When it rains, it pours." Although a clever copywriter got the credit, SSA is a complex inorganic salt used for its anticaking properties. It keeps a variety of powders free-flowing in the presence of moisture.
Regulated

Sodium Stearyl Fumarate is a derivative of fumaric acid. It is used in baked products as a dough conditioner to increase the rate of "aging."
Regulated

Sodium Stearyl-2-Lactylate is another derivative of lactic acid and has also been approved by the FDA as an emulsifier in leavened baked goods.
Cleared

Sodium Tartrate is a derivative of tartaric acid and is used both for its acidic qualities that enhance fruit flavors—especially orange, lemon, raspberry, and grape—and in candies for tartness.
GRAS

Sodium Tripolyphosphate (STP), no relation to the gasoline and oil treatment for automobiles, is used in foods for the wide variety of effects it can produce. One of its most important uses is as a sequestrant in alcoholic beverages to prevent haze. Others are as an antioxidant in fatty foods, and in curing solutions to retain moisture of meats.
GRAS

Sorbic Acid is originally obtained from berries of the mountain ash—rowanberry tree. It has antimicrobial activity against yeasts and molds (fungi). As a compound with calcium or potassium (calcium sorbate or potassium sorbate) it is used to prevent microbial rancidity in butter, margarine, mayonnaise, and salad dressings. It is also used in wine, bread, cakes and pastries, fruit and soft drinks, processed cheeses, and pickle products.

Sorbitan Monostearate, a product of reacting stearic acid and sorbitol, is usually used along with Polysorbate 60, 65, and 80 in whipped toppings, cake mixes, and other products that can benefit from improvement in generation of air bubbles during creaming or mixing. Air bubbles have a leavening effect and control grain size.
Cleared

Sorbitol Along with mannitol and propylene glycol, it belongs to a family of compounds known as polyhydric alcohols or polyols—sweet alcohols. They all have the ability to absorb and retain moisture (they

are *hygroscopic*) and are added to foods to keep them moist ("fresh").
GRAS

Stannouschloride is an inorganic compound used for its ability to retain the green color of asparagus packed in glass containers.
Cleared

Sucrose is a household or table sugar. It is also the most abundant free sugar found naturally and has been used since antiquity. Sucrose is composed of glucose and fructose and is produced by concentrating sugar cane or sugar beet juice. Sugar has the unique ability to impart density and "mouth feel" to soft drinks, which artificial sweeteners cannot do.
GRAS

Sugar Beet Extract is the concentrated residue obtained from sugar beets, used for its flavor-enhancing properties. The concentrate has had both sugar and glutamic acid removed and only a minimum of naturally occurring trace minerals remains.

T

Tagetes (Aztec Marigold) is a flavoring oil with an intense herbal scent obtained from the marigold, *Tagetes patula*.
Regulated

Tartaric Acid, an edible organic acid found widely in nature, is an acidulant with a strong tarty flavor. Used in jams, jellies, candies, it can aid in preventing rancidity and discoloration of foods.
GRAS

Tertiarybutylhydroquinone (TBHQ) Anything with a name like this ought to be banned on that basis

alone. Actually, it's been found to be quite harmless. It is an antioxidant that is almost always used together with BHA or BHT to retard rancidity in baked products.
Regulated

Tetrasodium Pyrophosphate (TSPP), an inorganic compound, has a variety of uses in food. One is moisture retention—especially in deboned meats. It is also used for its flavor-enhancing properties. This may be related to its moisture-holding ability. Another use is for preservation—preventing microbial growth on the surface of poultry. It is also used to aid in color retention in fish and shellfish, and in powdered beverage mixes to aid the dissolving of coffee and hot chocolate in water.
Regulated

Tocopherol, from the Greek meaning "to bear offspring," is also known as vitamin E. It occurs naturally in nuts and seeds, oils and fruits. It can be used in foods for its antioxidant (rancidity-retarding) ability, or as a nutrient supplement. Its offspring-bearing ability appears to be better in lab animals than in man.
GRAS

Tragacanth (See Gums.)
GRAS

Triacetin For colors to be used in foods, they must first be dissolved. Triacetin is a solvent used for its ability to suspend a variety of coloring additives. It can also be used for its antimicrobial activity in leavened products, piecrusts, and pastries.
GRAS

Tricalcium Phosphate (TCP), an inorganic compound (again, no carbon, no hydrogen), is often used to-

gether with the antioxidants propyl gallate and BHA
or BHT to prevent or retard rancidity in oils. For example, sunflower seeds, which have high levels of oil
and are likely candidates for oxidative rancidity,
usually contains a combination of TCP and propyl
gallate or BHT.
GRAS

Triethyl Citrate is a derivative of citric acid. It is a gen-
eral- or multi-purpose ingredient used primarily in
egg whites for its acidic preservation effects as well
as its ability to tie up trace metals that can cause off-
odors.
GRAS

Trihydroxybutyrophenone (THBP) Is it any wonder
that this was mercifully shortened to THBP? Although a synthetic organic compound produced by
the Kodak Company in 1963, it has been found safe
for use as an antioxidant. Most often it is used in
combination with BHA or BHT.
Regulated

Trisodium Citrate is a salt of citric acid which is used
in the powders that make fruit drinks. It aids in the
dissolving of these powders in water. These "salts"
have been used in foods for over 100 years. They are
general- or multi-purpose additives. They are used in
ice cream, sherbets, margarine, jellies and jams,
evaporated milk and nonalcoholic carbonated beverages.
GRAS

Turmeric is a natural flavoring obtained from the roots
of East Indian herbs of the ginger family—*Curcuma
longa*—has been used as a spice for hundreds of
years. Its intense yellow is made use of in meat products and salad dressings. (See Zeodory.)
GRAS

V

Vanillin, not to be confused with vanilla, a natural flavoring, is a synthetic substance with vanillalike odor and flavor.
GRAS

W

Whole Fish Protein Concentrate is obtained from such fish as hake, herring, and anchovy, and is the protein derived from treatment of the fish with alcohol. The alcohol extraction removes soluble solids. The concentrate is used to upgrade the protein content of foods.
Cleared

X

Xanthin is a carbohydrate gum obtained by bacterial fermentation of dextrose. The process was developed by the U.S. Department of Agriculture in its search for new uses for corn products.
Cleared

Xylitol is a five-carbon sugar (sucrose, fructose, and glucose are six-carbon sugars) about as sweet as sucrose. Chemically, its five-carbon structure provides it with resistance to bacterial attack. Thus, products made from it are well-preserved.
Cleared

Y

Yeast-Malt Sprout Extract used as a flavor enhancer in food, is a mixture of nutrients obtained from yeasts using enzymes from malt barley.
Cleared

Yellow Prussate of Soda, otherwise known as sodium ferrocyanide decahydrate, is an effective anticaking

agent as it causes sodium chloride (table salt) to produce star-shaped crystals rather than the usual cuboidal ones, which are less likely to cake.
Cleared

z

Zeodory is a flavoring obtained from the bark of an East Indian tree—*Curcuma zeodoaria*—used in ginger ale formulations and in bitters.
GRAS

11

Some Final Thoughts

Although, unfortunately, the term "food additive" raises a red flag with a number of people, the fact is that the three most widely used food additives are sugar (sucrose), salt (sodium chloride), and corn syrup. If to these are added citric acid, a multipurpose additive, the vegetable colors, the condiments pepper and mustard, and sodium bicarbonate or sodium acid carbonate (baking soda), they total more than 98 percent, by weight, of all the food additives used in the country today.

Now that's a fact. As the psychiatrist Robert Waelder noted in his trenchant book, *Progress and Revolution,* "Strongly held opinions often determine what kind of facts people are able or willing to perceive." It's another fact that strong but untutored opinion exists about food additives.

I do believe that three problems confront the American people today. These problems are, lack of information, conflicting information, and continuous exposure to misinformation. This, in the face of the most advanced communications technology and electronic wizardry for bringing information literally instantly from border to border and coast to coast. Given some reflection, however, it may just be that these problems are a *consequence* of the available technology—producing a babel of tongues, a form of mental pollution. It is something to dwell upon.

Underlying the above may be another problem, what Sigmund Freud called the "death wish" or "death drive." Any number of professionals concerned with mental health suggest, from time to time, that there appears to be a preoccupation or vicarious concern with death among segments of our population.

Confused by fear and misled by ignorance, people seem unwilling or unable to believe that they are, in fact, healthy—an especially ironic situation when you consider that the Surgeon General of the United States titled his annual report on the state of the nation's health, *Healthy People.*

In preference to feeling good about the high level of their health, the propensity is the other way: The total environment is seen as contributing to illness and/or disease. The grim scenario has it that we are sick or soon to be, and that there is no escape.

That's what large segments of the public believe. But that's not what another segment of the public, the medical-scientific community (with no special food supply of their own) knows to be the case.

The medical-scientific community knows that life expectancy at birth and longevity are both increasing substantially. They know too that the Infant Mortality Rate (IMR), a sensitive indicator of general environmental conditions, was 13 per 1,000 live births in 1979. And provisional data for 1980 indicate that the rate will be below 13. In 1960, the rate was 26. And before that it was higher each year all the way back to the beginning of the century. We know, too, that death rates for all ages are falling. Heart disease has been declining steadily since 1968. The incidence of stroke has dropped, and, if we deduct the number of cigarette-induced lung cancers from the overall cancer rate, cancer is seen to have reached a plateau. And infectious disease is almost a thing of the past. These are the easily verifiable facts of life.

Given the persistence of contrary opinion, a "death wish" does not seem so outlandish an explanation. What an awful thing to have, or to want to live with. The disparity between what large segments of the public prefer to believe and what the medical-scientific community know to be true, is not just wide, it is enormous.

The medical-scientific community knows that as a class of chemicals, food additives are the most tested, most scrutinized of any chemicals that come in contact with people. They know too that there has never been a single documented case of human illness or death attributed to them. Yet many people believe otherwise. It is well-nigh impossible to fathom what it will take to convince people of the safety of our food supply. What is it that nourishes the myth? A fantasy clearly opposed to people's best interests.

It is the profound discrepancy between these two sets of beliefs that so disturbs me and that has prompted me to write this book—an effort to relieve the problem of lack of appropriate information. The problems of conflicting information and continuous misinformation I see as a dilemma not easily undone. But they can be reduced if more professional scientists become concerned about the public's need to know—and to understand.

Of all the risk factors that bedevil us in this life, food additives have to be among the least troublesome—so small a problem that as a group their ill effects, if any, cannot be ascertained. Compared to the risk of death, injury, or illness from automobiles or motorcycles, climbing a mountain, spelunking, crossing a street on any day of the week, drinking liquid protein concentrate, using tampons, or being bitten by a rattlesnake in downtown Pittsburgh, food additives are a success story.

I suspect that G. K. Chesterton knew what he was talking about when he said, "The real trouble with this world of ours is not that it is an unreasonable world, nor even that it is a reasonable one. The commonest kind of trouble is that it is nearly reasonable, but not quite."